WEATHER

A Study Unit to Promote Critical and Creative Thinking

Written by Rebecca Stark
Illustrated by Karen Neulinger

WESTERN EDUCATIONAL ACTIV. LTD.
10929 101st St.
Edmonton, Alberta
T5H2S7
(403) 429-1086
FAX (403) 426-5102

ISBN 0-910857-79-2

© 1990 Educational Impressions, Inc., Hawthorne, NJ

EDUCATIONAL IMPRESSIONS, INC.
Hawthorne, NJ 07507

Table of Contents

Introduction

This learning unit was created to introduce youngsters to important weather concepts in a fun-filled way. An equally important objective of the unit is the development of crucial **critical and creative thinking skills**. The activities were specifically designed to encourage divergent thinking, flexibility of thought, fluent production of ideas, elaboration of details, and originality. They may be adjusted to suit your particular teaching style and/or the ability levels of the children. You might also want to rearrange some of the activities to coincide with weather conditions.

Rebecca Stark

Teacher Suggestions With Select Answers

Nice and Breezy (page 10)—Extended activity: Obtain a copy of the Beaufort Wind Scale, a system for measuring and classifying wind speeds. Have the students draw pictures to illustrate each of the 12 classifications. You might want to have the children work in 12 small groups. In that case you would assign each group a different number on the scale. Display the pictures around the room or on the bulletin board.

Where Did All the Water Go? (page 13)—Elicit from the youngsters that the water in the saucer and in the puddle evaporates. It goes into the air, but we cannot see it.

It's Raining; It's Pouring! (page 16)—Drizzle, snow, hail, sleet, and glaze are other forms of precipitation. If any of the children list dew or frost, elicit from them that they are not really forms of precipitation. (Refer to the activities on pages 36 and 37.)

What Can You Make of These? (page 20)—Encourage children to be creative and to stretch their imaginations with this and other open-ended activities. Give extra praise for unusual details, using more than one cloud per picture, etc.

Snowflakes (page 22)—1. D   2. B   3. A   4. C

Snow Helps in Many Ways (page 23)—A woodchuck is hibernating. Elicit from the youngsters the other ways in which snow benefits us. These are possible answers. It covers shrubs and other plants by keeping away the wind, ice, and cold. Igloos, which are made with blocks of snow, help keep the wind, ice, and cold away from people. Snow melts and changes to water; it fills our reservoirs, streams, and rivers. Melted snow goes into the soil and helps make the soil good for growing plants. Snow also provides a lot of fun for many people.

Oh, No! Snow! (page 25)—The dot-to-dot is a deer. Elicit from the children the various ways in which snow can be harmful. These are some possible answers. It makes walking and driving dangerous. It may make it difficult for animals to get around or to find food. Cars and other vehicles get stuck. Power lines can come down. If the snow melts too quickly, it can cause flooding. It causes extra work, such as shoveling sidewalks and driveways.

It's Snowing (page 28)—Answers will vary, but these are a few possibilities. Those who like the snow might include children who get a day off from school; skiers; people who like to ride sleds; auto body shop owners; people who sell shovels, snowblowers, etc.; people who don't have to drive in it; people in areas where there is a water shortage; and so on. Those who do not like the snow might include people who must drive to get to work; people who have to drive as part of their job (bus drivers, for example); police officers who have to go to the scenes of the accidents that might occur; people who have to take their dogs for a walk in the snow; people who have no one to shovel their walkways and/or driveway for them; and so on.

Snowperson Finger Puppets (page 29)—Depending upon the ability level of the children, you may have to cut out the puppets for them. This activity can be done as a group or the children can take turns reciting the poem. Elicit the verb changes from the youngsters.

Out of Order (page 30)—A. 4   B. 2   C. 1   D. 5   E. 6   F. 3

Snowflake Pattern (page 31)—Point out to the children that there are six sides to the figure. Explain that although no two snowflakes are exactly alike, they all have six sides. Extended activity: Take the children outside (or suggest that they ask their parents to do this with them the next time it snows). Try to catch a large snowflake on a sheet of black paper. Examine the flake through a pair of binoculars or a strong magnifying glass. See if you can see the design. The snow melts quickly! Work fast!

Change Rain to Hail (page 33)—rain—pain—pail—hail

Making Dew (page 34)—Water vapor from the air will condense on the glass. The outside of the glass will become wet. Elicit from the children that if the glass stays dry, not much water vapor is in the air.

Thunderstorm Alert (page 35)—Depending on the age and ability level of the children, you might want to tell them that sound travels at 1,120 feet (340 meters) per second.

Show Me the Way to Go Home (page 38)—The rule is "Don't Cross Any Lines." Elicit from the youngsters the ways in which they went about solving the maze (trial and error, guessing, tracing with their fingers before using the pencil, etc.).

Hurricane Warning (page 39)—It looks like the storm has passed to those in the calm of the "eye." But the area will soon be hit again by the other side of the hurricane.

Over the Rainbow (page 41)—Depending upon the age and ability level of the children, you might want to explain that although light looks clear, it is really made up of many colors. If you have a prism, illustrate how the prism breaks up the light into bands of colors: red, orange, yellow, green, blue, indigo, and violet. Explain that this band of color is called a spectrum.

Meteorologist (page 46)—The following words can be made from the letters in the word *meteorologist*. Some may be too difficult for your students. Use your judgment.

eel	grim	let	lost	moose	see	stir	tool
elm	grime	lie	lot	most	seem	stole	toot
emit	grist	liege	me	motor	seer	stool	tore
gem	gristle	lime	meet	or	set	store	torso
get	grit	list	melt	ore	silt	street	tot
girl	groom	lit	mere	reel	sit	tee	tree
gist	I	log	met	remit	sleet	test	trestle
glee	igloo	logo	meter	rim	slot	tie	trim
gloom	ire	loom	meteor	role	some	tier	trio
go	is	loose	mile	room	solo	to	trot
gore	isle	loot	mist	roost	spot	toe	
goose	it	looter	mitt	root	steel	tile	
got	leg	lore	mole	rot	stem	time	
greet	lest	lose	more	rote	stilt	too	

Measuring the Temperature (page 48)—Be sure youngsters understand the idea of a scale. Elicit from them that on these thermometers each line represents an increase or a decrease of 10 degrees. Explain that this is not always the case. Be sure that they understand that the circle (°) is the **symbol** for degrees. Elicit from them other symbols that they might know: +, −, =, and so on. Depending on the ability level of the children, you may want to tell them about centigrade and Fahrenheit.

My Weather Chart (page 49)—Be sure all the children understand the terms used on this chart. Explain any terms that seem unfamiliar to them.

Which Is Bigger? (page 51)—A. 2 3 4 1 5   B. 2 1 3 4 5   C. 5 4 3 2 1

Weather Word Search (page 58)—

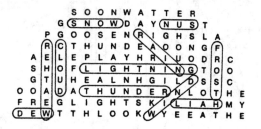

It's Raining Cats and Dogs (page 63)—The following are some idioms that may be familiar to the children: *to bite off more than you can chew, to have a trick up your sleeve, to be in a pickle, to keep your shirt on, to be in the dumps, to be in the doghouse, to jump out of one's skin, to break the ice,* and *to have cold feet.* If the children have trouble thinking of the idioms, give them a few and have them tell what they think they mean.

WHAT IS WEATHER?

Weather is the condition of the air around the earth. It is the condition at a certain time and a certain place. In other words, it is how WARM or COLD, WET or DRY, WINDY or CALM, and STORMY or CLEAR it is.

We get most of our heat from one very important source. Add something to the picture below that shows our main source of heat.

Now color the picture.

8

NAME THAT WIND

Wind is air in motion. The faster the air moves, the stronger the wind.

Winds are named after the direction FROM which they come. For example, a **west wind** comes **from the west**; an **east wind** comes **from the east**.

This "compass rose" shows the four basic directions: NORTH (N), SOUTH (S), EAST (E), and WEST (W).

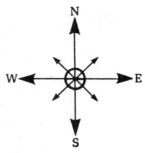

Now fill in the missing directions. Use N for north, S for south, E for east, and W for West.

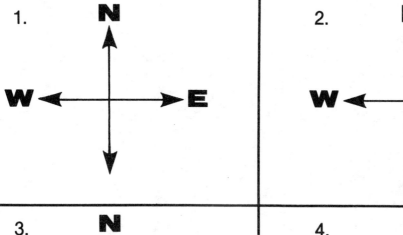

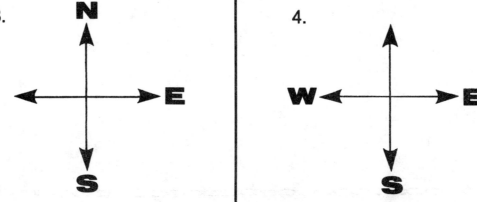

9

NICE AND BREEZY

When the wind blows at about 8 to 12 miles per hour, we call it a gentle breeze. Flags blow out. The small branches on trees move.

Draw a picture of something you like to do
when there is a gentle breeze.

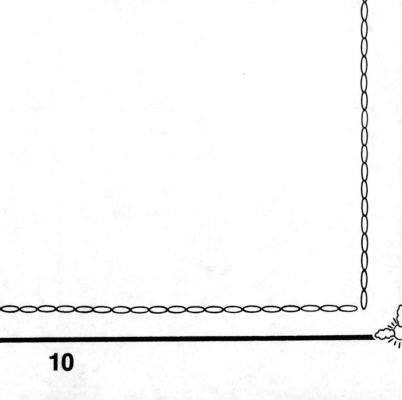

A BREEZY DAY

Many people enjoy flying kites on a breezy day. Study this picture for two minutes. Be sure to pay attention to small details as well as to the big picture.

Now turn the page over.

WHAT DO YOU REMEMBER?

Did you study the picture carefully? Now let's see how much you remember about it!

Answer these questions:

1. How many stars are on the kite? _____

2. How many bows are on the tail of the kite? _____

3. How many clouds are in the sky? _____

4. What animal is in the picture? _____

5. Is there a person in the picture? _____

6. What else do you remember about the picture? _____

_____

_____

_____

Now look at the picture again. See how well you remembered!

WHERE DID ALL THE WATER GO?

The sun heats the earth. As it does, it changes some of the water from the lakes, streams, and oceans into water vapor. Water vapor is what we call water if it is in the form of a gas. The big word for this process—the changing of water into water vapor—is called **evaporation**.

You can demonstrate evaporation. Put a small amount of water in a saucer. The next day check to see if anything happened to the water.

What do you think will happen? _____

_____

_____

Try the following experiment the next time it rains. Wait until it is no longer raining and the sun has come out. Find a small puddle. Take a piece of chalk and draw an outline around the puddle. Later that day or the next day see if anything has happened to the puddle.

What do you think will happen? _____

_____

_____

A CLOUD IS FORMED

Do you remember how the sun's heat changed some of the water from lakes, streams, and oceans into water vapor? Well, the water vapor makes the air warm and moist.

The warm, moist air rises. As it rises, it cools. If it cools enough, it becomes denser or more compact. If the water vapor becomes dense enough, we can see it. A cloud is formed! The big word for this process is **condensation**.

Steam is a kind of water vapor that is formed by boiling water. When you boil water in a tea kettle, the steam escapes and forms a small cloud.

Label the steam.
Color the steam white.
Color the teapot red.
Color its handle brown.
Color the stove green.

ALL KINDS OF CLOUDS

A cloud is made up of millions of droplets of water or ice crystals. There are many different kinds of clouds. Below are three main types.

CIRRUS CLOUDS are high clouds. They look like thin streaks or curls. They are made up of tiny ice crystals.

CUMULUS CLOUDS are fluffy, white clouds seen low in the sky. We often see them in fair weather. But sometimes they turn into storm clouds!

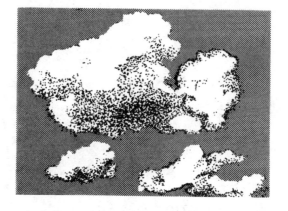

STRATUS CLOUDS are also low clouds. They look like a grey sheet. We sometimes get a fine drizzle from stratus clouds.

When clouds form very close to the ground, we call it **fog**.

Draw a picture using one of these cloud types.
Label the cloud cirrus, cumulus, stratus, or fog.

IT'S RAINING; IT'S POURING!

The millions of droplets in a cloud are always moving. As they move, they bump into each other. They join together and grow larger. Sometimes the droplets grow too large to float in the air as part of the cloud. They drop! A raindrop may be made up of a million or more droplets! The big word for the falling of water from the clouds is **precipitation**.

Rain is one kind of precipitation. What other kinds of precipitation can you think of?

_____

_____

_____

_____

_____

THE WATER CYCLE

The process of evaporation (water into water vapor), condensation (the formation of clouds), and precipitation (the falling of rain, snow, etc.) is called the Water Cycle. Study the picture for two minutes. Then turn to the next page.

The Water Cycle

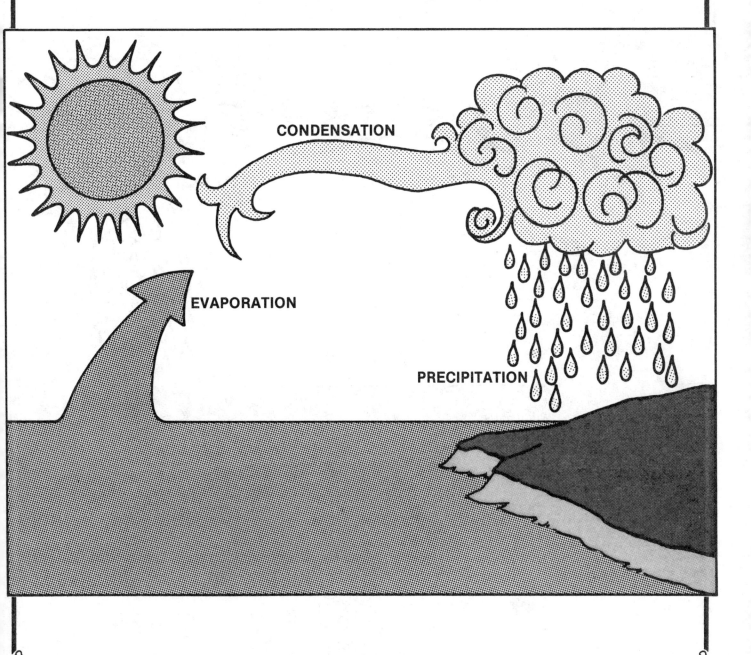

17

DO YOU REMEMBER?

See if you remember the process called the Water Cycle. Write these words in the correct places to show what is happening:

EVAPORATION   **CONDENSATION**   **PRECIPITATION**

© 1990 Educational Impressions, Inc.

18

WHAT DO YOU SEE?

Did you ever look at the fluffy, white clouds in the sky and imagine them to be other things? Use your imagination and tell what these "clouds" might be. They can be persons, animals, or objects—anything you want them to be!

WHAT CAN YOU MAKE OF THESE?

See what pictures you can make by using these clouds.

YOU'RE ALL WET!

Look at this picture. It is not raining. Why is Mary all wet? Write down as many ideas as you can think of. Stretch your imagination. Try to think of some unusual possibilities!

SNOWFLAKES

Sometimes the temperature in a cloud is below freezing. That cloud may be made of tiny ice crystals instead of water droplets. Precipitation will be in the form of snow.

Snowflakes appear as beautiful 6-sided crystals.

Look carefully at each set of snowflakes. For each, decide which one is **exactly** like the one on the left. Circle that picture.

1.  A.  B.  C.  D.

2. 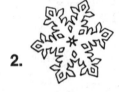 A. 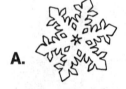 B.  C.  D.

3. 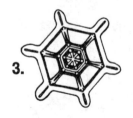 A.  B.  C. 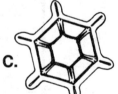 D.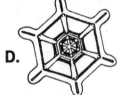

4.  A.  B.  C.  D.

SNOW HELPS IN MANY WAYS

Snow is helpful in many ways. One of the ways it helps is by covering the ground where animals are hibernating. The blanket of snow helps keep out the wind and the cold.

What are some other ways that snow is helpful?

_____

_____

_____

_____

_____

AN ICE HOUSE

An igloo is a kind of ice house. Eskimos sometimes build them to protect themselves from the cold arctic weather. Blocks of snow are used as bricks. Water is poured over the blocks of snow. The water quickly freezes and turns to ice. The ice holds the bricks of snow in place.

Look at the igloo at the right. Try to make it larger in the squares below.

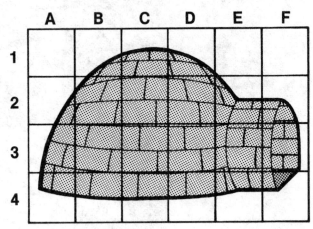

OH, NO! SNOW!

Although many people love snow, snow can cause problems. One thing that it does is make it more difficult for some animals to find food.

An animal is searching for food in the snow. Connect the dots to find out which animal it is. Be sure to keep the numbers in the correct order.

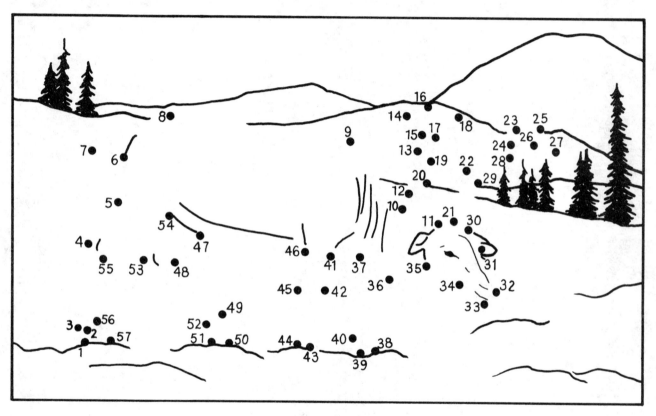

Color your picture.

What other problems might snow cause?

_____

_____

_____

_____

_____

_____

SNOW-TIME FUN

Snow can be a lot of fun! List all the ways you can think of that people can have fun in the snow. See if you can stretch your imagination and think of some unusual ideas.

Put a ★ next to your most original idea.

What do you like to do best when it snows?
On the next page, draw a picture to show what you like to do.

26

I LIKE THE SNOW

Draw a picture that shows what you like to do best when it snows.

IT'S SNOWING!

"It's snowing!" These words may greeted by joy or by worry. It depends upon who you are, what you do, when it is, where it is, and so on.

List all the people who might be happy that it is snowing. Tell why.

List all the people who might be upset that it is snowing. Tell why.

SNOWPERSON
FINGER PUPPETS

Use these patterns to make finger puppets. You will need a piece of cardboard for each puppet. Color your puppets. Add or change any details you wish.

Say the following poem several times. Each time you say it, change the action word. For example, you may change *jumping* to *dancing, running, hopping, skipping,* and so on. As you say the action word, act it out with your puppet. Stick your fingers through the holes as shown.

Jolly, jolly snowman (or lady) having fun at play.
JUMPING, JUMPING, JUMPING on this wintry day!

OUT OF ORDER

Look at the pictures below. They are out of order. Number the pictures so that they are in the correct order to tell a story.

A.

B.

___

C.

D.

___

___

E.

F.

___

___

SNOWFLAKE PATTERN

Make copies of this snowflake pattern. Cut out the snowflakes. You may vary your designs. But be sure that all six sides have the same pattern.

Hang the snowflakes around your classroom.

THE FIVE SENSES

Fill in the blanks to describe how rain and snow look, feel, sound, smell, and taste to you.

Snow

Snow sounds like _____

Snow tastes like _____

Snow feels like_____

Snow looks like _____

Snow smells like _____

Rain

Rain sounds like _____

Rain tastes like _____

Rain feels like _____

Rain looks like _____

Rain smells like _____

CHANGE RAIN TO HAIL!

Did you ever see a hailstone? A hailstone is made up of layers of ice and snow. If you cut a hailstone in half, it would look almost like an onion.

CROSS-SECTION OF HAILSTONE

CROSS-SECTION OF ONION

Most hailstones are about the size of a pebble. But some get to be as large as a softball!

Change RAIN to HAIL by changing one letter at a time. The clues on the left will help you.

rain

It hurts.   _____

You carry
things in it.   _____

hail

MAKING DEW

The amount of water vapor in the air is called **humidity**. Warm air can hold more water vapor than cold air. Therefore, when warm air turns colder, some of the vapor must turn back into water. We call that water **dew**. When the temperatures are below freezing, the water vapor changes to **frost** crystals instead of liquid water.

You can show how this happens. Get a glass of water. Add several ice cubes to the liquid. Put the glass in a warm room.

Finish the drawing to show what happens.

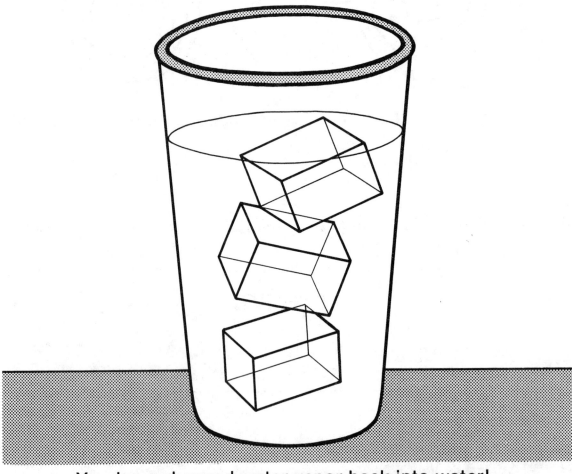

You have changed water vapor back into water!
You have made dew!

On another sheet of paper draw a picture of frost.

THUNDERSTORM ALERT!

Thunderstorms can be very exciting! **Lightning** is like a huge electric spark. It can take place within a cloud, between two clouds, or between a cloud and earth. The loud sound that follows the flash of lightning is called **thunder**. We see the lightning as soon as it happens. But it takes a while for the sound of the thunder to reach us. By knowing how long it takes, we can tell how far away the storm is. Suppose six seconds pass from the time you see the lightning to the time you hear the thunder. That would mean the storm is 1.3 miles (2 kilometers) away.

LIGHTNING SAFETY

Study these rules carefully before going on to the next activity.

Indoors:

1. Do not use the telephone.
2. Do not use electrical appliances.
3. Do not use a water faucet.
4. Do not fix a TV antenna.

Outdoors:

1. Lightning usually hits the highest places. Do not stand in an open field.
2. Do not stand under a tall tree.
3. Do not hold anything metal.
4. If you are in the water, get out at once.
5. A closed car is usually safe. If you are in a car, stay there.

DON'T DO THAT!
Outdoors

Think about the rules of lightning safety. Put an **X** through the dangerous things that people are doing in this picture.

DON'T DO THAT!
Indoors

Think about the rules of lightning safety. Put an **X** through the dangerous things that people are doing in this picture.

37

SHOW ME THE WAY TO GO HOME

A storm is heading Amy's way! Help Amy find her way home!

This kind of activity is called a MAZE. What rule must you follow when doing a maze?

_____

_____

START

HURRICANE WARNING!

Hurricanes are very violent storms! They start over the ocean and break up soon after hitting land. At one time, hurricanes were given girls' names. Today they are given both boys' and girls' names.

A hurricane has winds of 75 to over 150 miles per hour. It causes a great deal of damage.

Look at this diagram of a hurricane. The calm center is called an "eye."

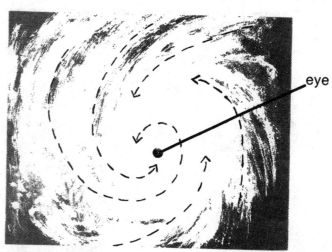

Now look at the picture at the left. The "eye" of the hurricane is passing over the island. What advice will you give the boy in the picture?

That was some storm! I'm glad it's over!

_____

_____

_____

_____

_____

_____

39

TORNADO!

A tornado has a whirling wind.
It is such a strong wind that it can lift cars
and throw them into the air. It can even throw
houses up into the air. Large trees are
pulled out of the ground by their roots. Luckily,
a tornado has a very narrow path.

Connect the dots to find out what a tornado cloud looks like. Be sure to keep the numbers in the correct order.

40

OVER THE RAINBOW

A rainbow is caused by the sun shining through rain, mist, or spray.

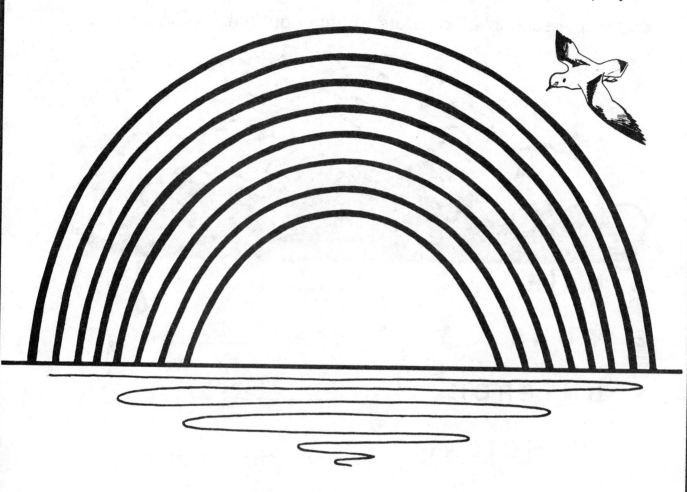

Color the rainbow from top to bottom:
- red
- orange
- yellow
- green
- blue
- indigo (a dark blue)
- violet (a reddish blue)

You can make your own rainbow! Stand with your back to the sun. Set a garden hose to a fine spray. A small rainbow should appear.

41

SCRAMBLED
WEATHER WORDS

Can you unscramble the letters to figure out these weather words?

1. **N A I R** _ _ _ _

2. **W O N S** _ _ _ _

3. **U N T E R D H** _ _ _ _ _ _ _

4. **I N T H G I L N G** _ _ _ _ _ _ _ _ _

5. **L O U C D** _ _ _ _ _

6. **U S N** _ _ _

7. **I N D W** _ _ _ _

8. **L I A H** _ _ _ _

9. **W E D** _ _ _

10. **R O F T S** _ _ _ _ _

HIDDEN WEATHER RHYME

There is a popular weather rhyme hidden in this puzzle. Take the correct path around the puzzle grid and you will find it!

START	I	T	'	S	R	A	I	N
G	.	T	H	E	O	L	D	I
N	R	I	N	G	.	THE END	M	N
I	O	N	S	S	I	N	A	G
R	U	O	P	S	'	T	I	.

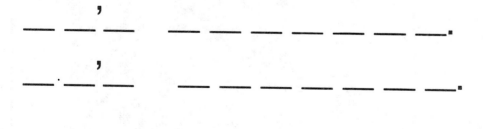

Write the weather rhyme here.

___ ___ , ___ ___ ___ ___ ___ .

___ . ___ , ___ ___ ___ ___ .

___ ___ ___ ___ ___

___ ___ ___ .

_____ FOR SALE!

Pretend that you have set up a sidewalk stand. You will sell things to people who pass by your stand. List all the things you will sell on a . . .

Rainy Day

Windy Day

Snowy Day

Hot Day

It is very windy! Parts of the letters have blown away. Figure out what the words are. Write the words on the lines.

1. WIND _____

2. PAIN _____

3. CLOUD _____

4. SNOW _____

5. HAIL _____

METEOROLOGIST

A person who studies the weather is called a **meteorologist**. That's a very big word! How many little words can you find hiding in it?

You may use a letter more than once only if it is in the big word more than once. Also, do not use the letter "s" unless it is part of the root of the word. In other words, you may use the word *list;* you may not use the words *lets* or *logs*. (You may, of course, use *let* and *log*.)

M-E-T-E-O-R-O-L-O-G-I-S-T

Write 25 words that you find in the word *meteorologist.*

_____ _____ _____ _____ _____

_____ _____ _____ _____ _____

_____ _____ _____ _____ _____

_____ _____ _____ _____ _____

_____ _____ _____ _____ _____

Can you think of any others?

_____ _____ _____ _____ _____

_____ _____ _____ _____ _____

_____ _____ _____ _____ _____

_____ _____ _____ _____ _____

P.S. There are over 100 little words hiding in *meteorologist!*

IT'S TOO PLAIN!

This raincoat is too plain! What details might you add to make it more interesting? Color the raincoat to show your improvements.

MEASURING THE TEMPERATURE

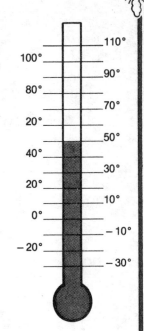

A thermometer is an instrument for measuring temperature. It is usually in the form of a glass tube with mercury or alcohol sealed inside. On the outside is a scale marked in degrees. The symbol for degrees is a tiny circle (°). Look at the thermometer at the right. It shows a temperature of 50°.

Look at the first thermometer. Find the line that shows 90°. Use a red crayon to color the tube up to that line.

Look at the second thermometer. Find the line that shows 40°. Use a blue crayon to color the tube up to that line.

Look at the third thermometer. Find the line that shows 0°. Use a green crayon to color the tube up to that line.

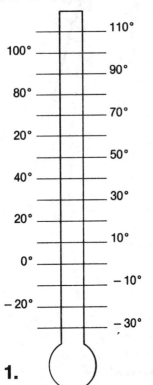

1.

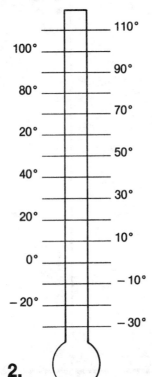

2.

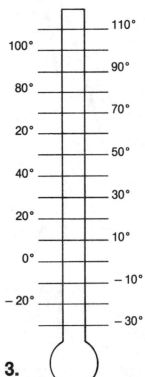

3.

MY WEATHER CHART

Use this chart to help you keep track of weather conditions for one week.

DAY OF WEEK	HIGH TEMPERATURE	LOW TEMPERATURE	SKY CONDITIONS	WIND CONDITIONS	TYPE OF PRECIPITATION

HIGH TEMPERATURE = the highest temperature reached on that day

LOW TEMPERATURE = the lowest temperature reached on that day

SKY CONDITIONS = fair (not cloudy or stormy), cloudy, partly cloudy, hazy

WIND CONDITIONS = calm, breezy, windy, stormy

TYPE OF PRECIPITATION = rain, freezing rain, snow, hail, mixed snow and rain

49

TODAY'S SPECIAL

Pretend that you own a small diner. What lunch specials might you offer to encourage customers to come? Complete the signs with words and/or pictures to show what you will offer.

It is
90°F (32°C)
and
sunny.

Today's Lunch Special

It is
18°F (-8°C).
It is
starting
to snow.

Today's Lunch Special

WHICH IS BIGGER?

What is meant by "biggest"? What is meant by "smallest"? For each set of pictures you will have to judge the sizes of the objects. For each line write #1 in the box under the smallest object in the set; write #2 in the box under the next smallest object in the set; and so on. The biggest object in each set should have the number 5 in the box.

WEATHER POEMS

Quatrains are 4-lined poems. Can you complete each of these quatrains? Choose from the following list of words:

white   ice   sky   pane   window

Jack Frost has been here.

How do I know?

He left me a picture

On my bedroom _____.

I like the cold.

I think it's nice.

It's fun to skate

On the smooth _____.

I sit in the parlor

Just watching the rain.

It makes a soft patter

On my window _____.

If I could only learn to fly,

I'd go way high up in the _____.

When I got there, I think I might

Float like a cloud, so fluffy and _____.

Now write your own 4-lined poem with a weather theme.

BIRTHDAY FUN

When is your birthday? _____

Where do you live? _____

Think about what the weather is usually like on your birthday. Pretend you are having an outdoor birthday party. Draw a picture to show what you and your guests might do.

A SHAPE POEM

A shape poem is written in the shape of the topic of the poem. Shape poems are sometimes called concrete poems. A shape poem does not have to rhyme.

Below are two examples of shape poems about boots.

BOOTS

Some shape poems give a thought about the subject.

Some shape poems use only descriptive words.

I love to splish splash in puddles when I wear my shiny new boots. shiny new boots.

Shiny, hard to get on, good for splashing in puddles, hard to get off, BOOTS.

On the next page you will be asked to create your own shape poem.

MY SHAPE POEM

Now write your own shape poem. Choose a weather theme. A few ideas are listed below. You may use one of these ideas or you may choose an idea of your own.

umbrella   cloud   snowflake   raincoat

My topic: _____

Write down words that describe your topic. Also include words that tell how you feel about the topic.

_____   _____   _____

_____   _____   _____

_____   _____   _____

_____   _____   _____

Use the list of words to help you write a shape poem.

_____ (Title)

TODAY'S WEATHER

Each of these children lives in a different place. Today's weather is different in each of those places.

Match each child's picture with the correct scene.

It is hot
and sunny
where
Jane lives.

It is warm
and rainy
where
Jill lives.

It is cold
and snowy
where
Bill lives.

What is today's weather like where you live?

_____

_____

DRESS FOR THE WEATHER

Each of the children described on the previous page wants to go outside to play. For each child, draw the proper outdoor clothing. Then draw a picture to show what you would wear if you went outside to play today.

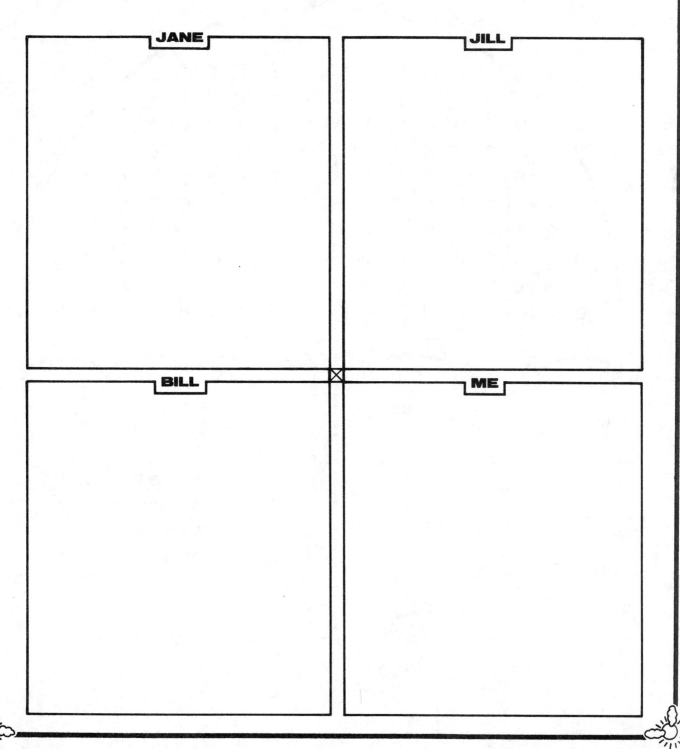

JANE

JILL

BILL

ME

WEATHER WORD SEARCH

See if you can find the words listed below. You may go in a straight line in any direction:

```
        S O O N W A T T E R
      G S N O W D A Y N U S T
    P G O O S E N R I G H S L A
    R C T H U N D E A O O N G F
  A E L E P L A Y H R I U O D R C
  S H O F L I G H T N I N G T O O
  G A U H E A L N H G I L D S S C
O O T D A T H U N D E R N L O T H E
F R A G L I G H T S K I L I A H M Y
D E W T T H L O O K W Y E E A T H E
```

cloud   snow

dew   sun

frost   thunder

hail   weather

lightning   wind

rain

Circle the words
as you find them.
One has been
done for you.

58

WHAT'S YOUR CLIMATE?

Weather is the condition of an area at a certain time. Climate is the weather of that area over a long period of time.

Scientists classify climates by how hot it gets and how much rain falls.

TEMPERATURE
- TORRID (very hot)
- TEMPERATE
- FRIGID (very cold)

* RAINFALL
- WET
- HUMID (a lot of moisture in air)
- SUBHUMID
- SEMI-ARID
- ARID (dry)

*Used especially to further describe torrid and temperate climates.

How would you describe the climate where you live?

_____

_____

_____

How would you describe the climate at the North Pole?

_____

_____

How would you describe the perfect climate?

_____

_____

_____

A CHANGE OF SEASONS

In some parts of the world, the weather stays the same all year long. But in many parts of the world, there is a change in seasons.

| January | May | July | October |

Can you name the four seasons? Write them on the lines below. Under the name of each season, write a list of words associated with that season.

_____   _____   _____   _____

Would you like to live in a place where the weather is always the same? Or would you rather live in a place where the weather changes with the seasons? Tell which you would prefer and why.

_____

_____

COLOR-CODED PICTURE

This picture is to be colored in according to a letter code. The letter "a" will stand for the color blue. That means you will fill in all the spaces marked "a" with blue. Let each of the other letters stand for a different color. Then color in the picture according to your code.

a = blue

b =

c =

d =

e =

What season is it in your picture? Describe the weather.

_____

_____

MY FAVORITE WEATHER

Describe your favorite weather.

_____

_____

_____

_____

_____

Now draw a picture that shows why it is your favorite weather.

IT'S RAINING CATS AND DOGS

"It's raining cats and dogs" is an idiom. That means that the words do not mean exactly what they say. What does "It's raining cats and dogs" really mean?

In each sentence the idiom is in **bold**. Tell what each really means.

1. I wanted to tell her, but I **held my tongue.**

_____

2. She **cried her eyes out.**

_____

3. Please **lend me a hand.**

_____

4. Everything I tell her **goes in one ear and out the other.**

_____

Can you think of any other idioms?

_____   _____

_____   _____

Choose an idiom. Pretend that the words really do mean exactly what they say. Draw a funny picture to illustrate the idiom. Use a separate sheet of paper.

A WORDLESS STORY

"Read" *The Snowman* by Raymond Briggs.* It is a story without words. In pictures only it "tells" of a little boy whose snowman comes to life.

Create your own wordless story with a weather theme. Use this sheet to sketch your ideas. Then redraw your story on other paper.

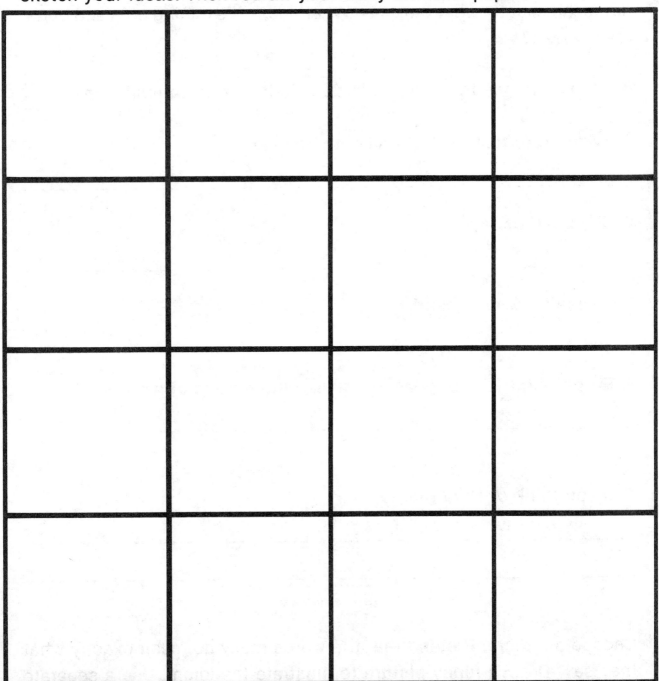

*Raymond Briggs, *The Snowman*. New York: Random House, 1978.